Carbohydrates

What You Need to Know

Written for
The American Dietetic Association
by Marsha Hudnall, MS, RD

CHRONIME[illegible] PUBL

Library of Congress Cataloging-in-Publication Data

Carbohydrates / The American Dietetic Association
p. cm.
Includes index.

ISBN 1-56561-143-8; $6.95

Edited by: Jeff Braun
Cover Design: Terry Dugan Design
Text Design & Production: David Enyeart
Art/Production Manager: Claire Lewis

Printed in the United States of America

Published by
Chronimed Publishing
P.O. Box 59032
Minneapolis, MN 55459-0032

10 9 8 7 6 5 4 3 2 1

Notice: Consult a Health Care Professional Because individual cases and needs vary, readers are advised to seek the guidance of a licensed physician, registered dietitian, or other health care professional before making changes in their prescribed health care regimens. This book is intended for informational purposes only and is not for use as an alternative to appropriate medical care. While every effort has been made to ensure that the information is the most current available, new research findings, being released with increasing frequency, may invalidate some data.

Carbohydrates
What You Need to Know

Written for The American Dietetic Association by
Marsha Hudnall, MS, RD
Nutrition, Health & Fitness Communications, Ludlow, Vermont

The American Dietetic Association Reviewers:
Judi Adams, MS, RD
Wheat Foods Council
Englewood, Colorado

Jackie Newgent, RD
Nutrition Communications Consultant
New York, New York

Alta Engstrom, RD
General Mills
Minneapolis, Minnesota

Technical Editor:
Raeanne Sutz Sarazen, RD
The American Dietetic Association
Chicago, Illinois

CHRONIMED PUBLISHING

THE AMERICAN DIETETIC ASSOCIATION is the largest group of food and health professionals in the world. As the advocate of the profession, the ADA serves the public by promoting optimal nutrition, health, and well-being.

For expert answers to your nutrition questions, call the ADA/ National Center for Nutrition and Dietetics Hot Line at (900) 225-5267. To listen to recorded messages or obtain a referral to a registered dietitian (RD) in your area, call (800) 366-1655. Visit the ADA's Website at www.eatright.org.

Contents

Introduction

ARE CARBOHYDRATES FATTENING? If you believe a few of today's popular diet books, carbohydrates are the dietary culprit responsible for the obesity epidemic sweeping America today.

The rationale behind this claim sounds impressive. The problem is, it's not accurate. It's just a myth—one that has taken the lead among many long-held misconceptions about carbohydrates. How many of these other carbohydrate myths have you heard?

- Bread and pasta are fattening.
- Sugar makes kids hyperactive.
- Brown sugar and honey are more nutritious than white sugar.
- High-protein, low-carbohydrate diets are best for fitness and weight loss.
- Refined carbohydrates make you fat.
- Sugar causes hypoglycemia and diabetes.
- Take fiber supplements to make sure you get enough fiber.

Many of these myths about carbohydrates defy national nutrition advice to eat plenty of carbohydrate-rich foods as the foundation of a healthy diet. Because they often sound logical, such misconceptions cause a lot of confusion about what to eat to promote health. People who believe they must forgo carbohydrate-rich foods to eat healthfully may just forgo eating healthfully instead.

This book sets the record straight about these carbohydrate myths and many others. It will help you understand the role of carbohydrates in diet and health and guide you in planning a healthful diet for you and your family.

Are You In-the-Know About Carbohydrates?

This quick quiz gives you a preview of some of the carbohydrate facts you'll learn about in this book. Take it before reading further to see how carbohydrate-savvy you are.

Q. Where do most of the carbohydrates in your diet come from?

A. Most of the carbohydrates you eat come from plant-based foods such as grains (bread, cereal, rice, and pasta), vegetables, fruits, and legumes (beans and peas). Milk and other dairy products such as yogurt and ice cream are the only animal foods that contain a significant amount of carbohydrates.

Q. How many servings of grain foods, vegetables, and fruits should you eat each day?

A. The Food Guide Pyramid recommends 6 to 11 servings of grain foods, 3 to 5 servings of vegetables and 2 to 4 servings of fruit (see page 53 for more information about the Food Guide Pyramid).

Q. Are bread and other starchy foods fattening?

A. No. Actually, no food in itself is fattening. You gain weight when you eat more food than you need. And remember, it's what you add to starchy foods that drives up their calorie content, such as margarine or butter on bread, or cream or cheese sauces on pasta. For more information about weight management and carbohydrates, see Chapter 5—Carbohydrates: Fueling Up for a Healthy Weight & Physical Fitness.

Q. Is sugar a carbohydrate?

A. Yes. Along with starch and fiber, sugars belong to a nutrient category called carbohydrates. See page 4 for more information.

Q. How much fiber should you eat each day?

A. Adults need 20 to 35 grams of fiber a day to gain the health benefits linked to a fiber-rich diet. That's twice the amount of fiber Americans typically eat. For more information about fiber, see Chapter 2—Dietary Fiber: The Rough Stuff.

Q. Can you get the same health benefits from a fiber supplement as you do from a high-fiber diet?

A. No. The health benefits linked to a high-fiber diet may also come from other components in fiber-rich foods, not just fiber alone. These foods contain a wealth of other disease-fighting substances that you won't find in a fiber supplement. For good nutrition, choose food first.

Q. Do carbohydrates like sugar cause kids to be hyperactive?

A. No. Studies show carbohydrates, including sugar, may actually calm kids (and adults). Carbohydrates may help produce serotonin, a brain chemical that helps calm and soothe jangled nerves. The calming effects of carbohydrates may go unnoticed, however, depending on the situation, such as an exciting birthday party. For more information, see page 46.

Q. Are honey, brown sugar, and maple syrup more nutritious than white sugar?

A. No. There isn't a significant nutritional difference between any of them. So, choose whichever one you like! Like all carbohydrates, they are broken down by the body and eventually converted into blood glucose (blood sugar). See "For Good Health, Which Sugar Should You Choose?" on page 2.

Chapter One

"Basic Training" in Carbohydrates

SUGARS, STARCHES, AND FIBER are the three types of carbohydrates that make up most of the carbohydrates we eat. Because of their chemical structure, sugars are referred to as *simple carbohydrates* or simple sugars and starches and fiber as *complex carbohydrates.*

All carbohydrates, whether simple or complex, are made up of three elements: carbon, hydrogen and oxygen. That's why they are called carbohydrates. *Carbo-* means carbon; *-hydrate* means water, or H_2O, which is made up of hydrogen and oxygen.

To make the different types of carbohydrates, these elements—carbon, hydrogen, and oxygen—are first arranged in a single *sugar unit*. Sugars, which have a *simple* structure, are made of one or two sugar units; and starches and fiber, which have a more *complex* structure, have many sugar units.

Simple Carbohydrates: Sweet & Not-So-Sweet

Sugars are referred to as simple carbohydrates. The term sugars, however, may need a little explaining. Table sugar, technically called sucrose, is just one of a number of different types of sugars in food.

Simple carbohydrates are classified according to how many sugar units they have. Monosaccharides contain one sugar unit. Disaccharides contain two sugar units. (*Mono-* means one, *di-* means two, and *-saccharide* means sugar.) The three types of monosaccharides are fructose, galactose, and glucose. When two monosaccharides are joined together chemically, they become disaccharides:

monosaccharide	+	monosaccharide	=	dissaccharide
glucose	+	fructose	=	sucrose
glucose	+	galactose	=	lactose
glucose	+	glucose	=	maltose

Fructose, which is found naturally in honey and fruit, is the sweetest of all the sugars. It has almost twice the sweetening power of sucrose, which is naturally found in some fruits and vegetables. Not all sugars are sweet, however. For example, lactose, which is found naturally in milk, is only slightly sweet.

Some simple carbohydrates, mainly sucrose and fructose, are used in food processing to make baked goods, frozen desserts, fruit juice drinks, jams, pudding mixes, soft drinks, and more.

And don't forget your own kitchen. In one form or another, you probably use sugar: white sugar, brown sugar, corn syrup, molasses, honey, jam, jelly or syrup.

Choosing sweetened foods or adding sugars to food is fine. In fact, sugar may add appeal to nutritious foods you otherwise avoid. However, important nutrients may be deficient in your diet if high-sugar, low-nutrient foods crowd out more nutritious options. For example, when soft drinks, rather than milk or fruit juice, frequently accompany a meal. Follow the recommendations of the Food Guide Pyramid for your best guide to healthful eating. See page 53 for more information on the Food Guide Pyramid.

For Good Health, Which Sugar Should You Choose?

Do you wonder whether you would benefit nutritionally if you used honey instead of table sugar (also known as sucrose) to sweeten your tea? Or if brown and raw sugars, molasses, or fructose rank higher on the nutritional scale than table sugar? Fact: There isn't a lot of nutritional difference between any of them. Let taste be your guide—choose what you like!

Table sugar, or sucrose, is also called refined sugar. It is made by extracting the juice from sugar cane stalks or the root of the sugar beet, then drying it to harvest the sugar crystals.

Brown sugar is merely white sugar crystals that are flavored with molasses. You might think of it as "value-added" table sugar because of the extra flavor it gives to foods.

Fructose is one of the simple sugars that makes up table sugar (sucrose = fructose + glucose). It's found naturally in fruits and is also added to some foods, either as crystalline fructose or as high-fructose corn syrup (HFCS).

Honey is made by bees from the nectar of a flower. Ounce for ounce, the nutrient content of honey and table sugar is about the same. But honey does differ from table sugar in several ways. Honey contains almost 20 percent water. That is important in cooking; using honey in a recipe that calls for sugar will add more liquid. Honey is higher in calories than table sugar but it's sweeter, so less can be used. The biggest difference: Honey shouldn't be given to infants. Honey may contain spores of Clostridium botulinum, a microorganism that leads to botulism. The spores can survive in the intestinal tracts of infants, where they can produce a deadly toxin. Older children and adults can safely enjoy honey because the spores can't produce the toxin.

Molasses is the thick, syrupy liquid that's left after sugar cane or sugar beets have been processed to make table sugar. Its distinctive difference is its flavor and color.

Raw sugar is the coarse granulated solid sugar left when sugar cane juice evaporates during the processing of cane sugar (table sugar). Turbinado sugar is raw sugar that has been refined to remove impurities usually present in raw sugar.

More Than Just a Sweet Taste

Sugars are prized for their sweet taste, but they also contribute aroma, texture, color, and body to foods—all of which add to our enjoyment. Sugars can also act as a preservative in food.

In baked goods:

- Sugars help develop the brown, flavorful crust in baked goods that's so important to our enjoyment. The browning is called caramelization.
- Sugars also help incorporate, retain, and stabilize air for light textures.
- Sugars act as food for yeast in yeast breads, and help all breads and many baked goods stay fresh longer because they hold moisture.
- In baked goods with limited moisture, such as cookies, sugars contribute to the crisp texture we enjoy.

In beverages:

- Sugars supply body, or "mouthfeel." Sugars also enhance and balance flavors.

In candies:

- Sugars are key to texture and flavor, whether they be hard candies or creamy fudge.

In jams, jellies, and syrups:

- Sugars bind water and preserve against the growth of yeast and molds.

Complex Carbohydrates: Digestible & Non-digestible

Starches and fibers are referred to as complex carbohydrates. Technically, they are called polysaccharides. *Poly-* means many and *-saccharide* means sugar. Yes, starches and fiber are made up of many sugar units. In essence, they are just longer chains of the sugar units found in simple carbohydrates.

If complex carbohydrates are just longer chains of simple carbohydrates, then why don't starchy foods, such as potatoes, pasta, breads, and beans, taste sweet like many simple carbohydrates? The answer is its larger structure. The larger molecules don't fit on the tongue's taste bud receptors. Starchy foods, however, can be broken down into smaller sugar units that you can taste as sweet. For example, keep a cracker in your mouth for a while. The digestive enzymes in your saliva will begin to break down the chains of sugar units. You'll notice a sweet taste as they become smaller.

Glycogen is a form of complex carbohydrate that is stored in your body, specifically in muscle tissue and the liver. It is actually made in the body from the carbohydrate-rich foods you eat and serves as the body's storage form of energy. Glycogen is used to power working muscles. See Fueling Physical Activity on page 36 for more information.

During digestion all carbohydrates except fiber are broken down into sugars. With fiber, your body's digestive enzymes cannot break down the long chains of sugar units that make up fiber in order to absorb it into your bloodstream. Instead, fiber passes through your digestive system unchanged and is excreted. This accounts for its special health benefits. Fiber is found naturally in whole-grain products, fruits, vegetables, beans, and peas.

The foundation of a healthy diet features foods rich in complex carbohydrates—grain foods such as bread, cereal, rice, and pasta, and vegetables, fruits, and legumes. In addition to supplying carbohydrates, these foods are the only source of fiber in the diet. Dairy products, meat, fish, and poultry don't contain fiber. What's more, many grain foods and most vegetables and fruits are low in fat. They also contain a wealth of other nutrients, such as vitamins, minerals, and phytochemicals (plant chemicals) that play important roles in health (see page 40).

Did you know the type of carbohydrate found in a plant may change as it matures? As fruit ripens, it gains a sweet taste as its starches change to sugars. Fruits often need a period of ripening before they are ready to eat. In contrast, many vegetables are sweeter when they are young. The sugars in peas, carrots, and corn turn to starch as they grow older. That's why good cooks choose young, fresh vegetables and use them quickly.

Carbohydrates: Essential to Life

Carbohydrates rank as the most common organic chemical on earth. Carbohydrates come from the radiant energy of the sun. The green leaves of plants trap this energy and transform it into the simple sugar glucose through photosynthesis—to the tune of about 100 billion tons a year. The glucose the plants don't use to meet their own energy or structural needs is then stored as sugars or complex carbohydrates.

Carbohydrates serve as the main energy source for animals and humans as well. Carbohydrates supply most of the energy to keep your heart beating, for breathing, and other body functions such as walking and running. It's easy to see how essential carbohydrates are to life.

When you eat carbohydrate-rich foods, whether foods high in simple or complex carbohydrates, they must be broken down into their simplest form before they can be absorbed. Enzymes in the digestive tract break down the carbohydrates you eat into sugars. There is one exception, however. Your digestive enzymes don't work on fiber. As a result, fiber passes through your digestive tract to provide its own special health benefits.

Are All Carbohydrate Calories Created Equal?

Here are the stats:

One gram of sugar provides 4 calories.

One gram of starch provides 4 calories.

One gram of fiber provides 0 calories.

In terms of calories, it doesn't make a difference whether calories in food come from sugars or starches. And since fiber isn't digested, it doesn't provide any calories.

Note: Calories are a measure of how much energy a food contains. One gram (unit of measurement) is about the weight of a regular paper clip or raisin.

Glucose—The Key Carbohydrate

Glucose is the main form of carbohydrate used for energy in the body. In fact, in the blood, glucose is also known as blood sugar.

In healthy people, a minimum amount of glucose is always in the blood. After eating, when carbohydrates are digested and then absorbed, the glucose, or blood sugar, level rises. It gradually drops again as tissues, with the help of the hormone insulin take glucose from the blood to use as energy. Glucose is also stored in the muscles and liver as glycogen. Some glucose may be converted to body fat if you consume more calories than you need. However, this is true whether the excess calories come from carbohydrate, fat, protein, or alcohol.

Because glucose is the major source of energy for nervous tissue (including the brain) and the lungs, the body has two back-up systems for producing it. The first is to convert the carbohydrate stored as glycogen back into glucose. This helps keep your blood sugar level within a normal range between meals.

When you don't eat enough carbohydrates, however, the body uses a back-up system called gluconeogenesis, or "making new glucose." The body uses parts of the fat or protein you eat to make glucose.

Just because this back-up system exists, it doesn't mean it's ideal. You can think of it as a way your body helps ensure that you survive for a time, even when you don't—or can't—feed yourself properly.

The drawback to the system is that without carbohydrates, your body doesn't function as well as it should. You start to lose body water and sodium, which accounts for the rapid loss of weight people usually experience on low carbohydrate fad diets. (By the way, it's water weight that quickly returns once you resume eating carbohydrates.)

Along with sodium, you also lose potassium. This can produce a feeling of weakness and sometimes muscle cramping. Further, the body often starts to break down body protein, which also contributes to weight loss. But more importantly, it can lead to a loss of muscle, something that's not desirable either for body tone or health.

Another consequence of a diet low in carbohydrates is the build-up of byproducts called ketones, which form when fat isn't completely metabolized. Fat is used as an energy source when carbohydrates aren't available. But carbohydrates are needed to fully metabolize fat. Without carbohydrates, ketones build up. This can lead to dehydration and further fatigue. In extreme cases, as may occur when diabetes goes undiagnosed, it can result in coma and even death.

The symptoms of ketone build-up can be quickly reversed by eating carbohydrates. The whole sequence of events, however, serves to show simply how important carbohydrates are to health. In people with diabetes, who can't utilize carbohydrates properly, the treatment is more complicated. Medication such as insulin may be needed.

Calories = Energy

Just how much energy do you get from carbohydrates? You're probably more familiar with the term calories when talking about the energy we get from foods. Carbohydrates are one of the lowest-calorie nutrients.

Carbohydrates* = 4 calories/gram
Protein = 4 calories/gram
Alcohol = 7 calories/gram
Fat = 9 calories/gram

*Not including fiber, which contains zero calories.

Chapter Two

Dietary Fiber

The Rough Stuff

IN THE 1800s, fiber was already a household name, although it wasn't actually called fiber. Most people referred to it as roughage, and realized it played an important role in health. But even though our awareness of the benefits of fiber predated the current interest in health and nutrition, Americans have been eating less and less fiber since the turn of the 20th century.

That's a problem because research continues to point to the health benefits of a diet high in fiber. By eating a low-fat diet that contains plenty of fiber, you may lower your chances of developing a multitude of health-related problems, including heart disease, some types of cancer, stroke, diabetes, diverticulosis, hemorrhoids, and more. Also, low-fat, high-fiber diets offer help if you are trying to lose weight.

Dietary Fibers—Not the Same but Equally Valuable

Dietary fiber in general refers to the complex carbohydrates in fruits, vegetables, grains, nuts, and legumes that can't be broken down by human digestive enzymes. Meats and dairy products do not contain fiber.

There are two basic types of fiber: insoluble and soluble. Most fiber-containing foods feature both. You can often tell the type of fiber that predominates in a specific food by its texture.

For example, the tough, chewy nature of wheat kernels, popcorn, apple skin, and nuts are due to the insoluble fiber they contain. While oats and beans are rich in both soluble and insoluble fiber, when you cook them, it's the soluble fiber you notice. The mushy texture of cooked oats and beans is due to soluble fiber, which soaks up water to give these foods their characteristic feel.

The Health Benefits of Fiber

Insoluble and Soluble Fiber Each type of fiber has its own individual effects on body function. Both types of fiber, however, may be important for waistline watchers. High-fiber foods create satiety. In other words, they create a feeling of fullness. They also take longer to chew, which may slow you down so you don't eat as much at a meal. What's more, fiber-rich foods are usually low in fat but contain plenty of other nutrients that are key to a healthful diet. And remember, fiber doesn't contain any calories. For dieters, foods that are rich in nutrients—not calories—make the best sense.

Insoluble Fiber The greatest impact of insoluble fiber appears to be its action in the large intestine, or colon. Because it is not digested, large amounts of insoluble fiber add bulk to the contents of the colon. It also acts to pull water into the large intestine, kind of like a sponge. A larger, softer stool is the result—one that exerts less pressure on the colon walls and is eliminated more quickly.

Less pressure on the colon walls reduces the risk of diverticulosis (small herniations in the colon wall that may become inflamed) and hemorrhoids, health problems that plague many Americans as they grow older.

Insoluble fiber may help reduce risk for more life-threatening problems as well. Large amounts of insoluble fibers both dilute the concentration of potential carcinogens (cancer-producing agents) in the stool and reduce the amount of time the colon wall is exposed to the substances. Further, insoluble fibers help keep the pH balance of the intestinal contents at a level that reduces microbial activity that can produce carcinogens. As part of a diet low in fat, the ultimate impact of a diet rich in insoluble fiber may be a reduced risk for colon and rectal cancers.

Fiber Tip: Bran Facts

When people think of fiber, they often think of bran. Bran is the coarse, outer layer of the whole-grain kernel. The kernel itself contains fiber as well as some starch, protein and a small amount of fat. The amount of insoluble and soluble fiber depends on the type of grain. For example, oat bran is rich in both soluble and insoluble fiber whereas wheat bran is predominantly insoluble fiber.

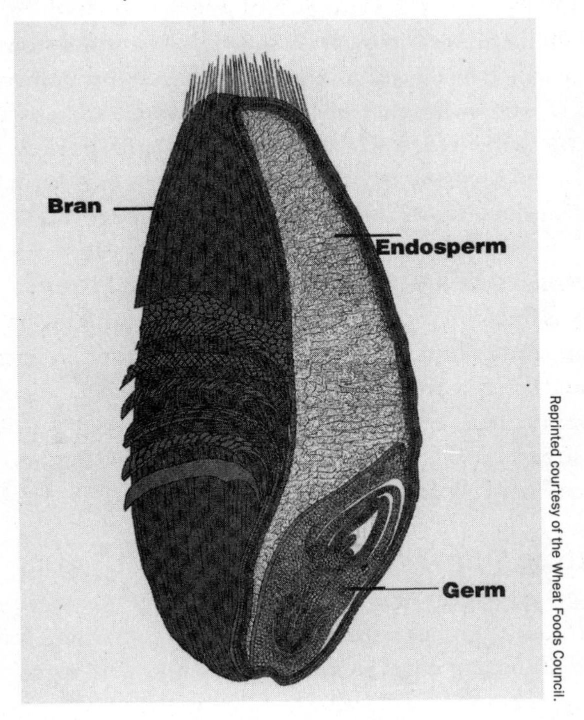

Reprinted courtesy of the Wheat Foods Council.

Whole grain = the entire edible part of any grain kernel—corn, wheat, oats, barley, rice, etc.

Soluble Fiber Like insoluble fiber, soluble fiber helps to keep you healthy. Soluble fiber may help reduce blood cholesterol levels, thereby reducing risk of heart disease. And a diet rich in soluble fiber may also help control the rise in blood sugar levels following a meal, which is important for people with diabetes.

Research shows that soluble fiber may reduce blood cholesterol by preventing the reabsorption of bile acids into the body. It works like this: Bile acids, which are made from cholesterol, are released into the small intestine to help digest foods. As food is digested and absorbed, the bile acids are reabsorbed into the body.

Soluble fiber, however, appears to tie up bile acids in the small intestine and thereby block their reabsorption. As a result, more cholesterol in the body is used to make more bile acids, causing the level of blood cholesterol in the body to drop.

Soluble fibers also increase the viscosity of the stomach contents, which may result in the stomach contents emptying more slowly. Therefore, the subsequent digestion and absorption of glucose (which food is broken down into) may be slower as well. Slower absorption of glucose may help people with diabetes control the level of their blood sugar (see page 41).

Which Foods Are Good Sources of Fiber?

Fiber is found in plant foods only—fruits, vegetables, beans, peas, whole-grains, nuts, and seeds. These foods contain both insoluble and soluble fibers—just different amounts in a specific food. In Appendix II, you will find the fiber content of fiber-rich foods. Because there are different types of fiber in foods, choose a variety of fiber-rich foods daily.

Getting Your Fiber Fill

For good health, adults need 20 to 35 grams of fiber daily. Children often need less because fiber-containing foods tend to be filling and may interfere with a child's ability to consume enough calories. Children need plenty of calories for growth. The simple formula—age + 5 grams—guides you in determining how much fiber kids over the age of 2 and under 20 need. For example, a 10 year old needs about 15 grams of fiber a day.

How does your diet measure up when it comes to fiber? Most Americans get half the recommended amount. If you're not currently getting enough, choose fiber-rich foods over fiber supplements. Fiber supplements don't contain all the other health-promoting substances such as vitamins, minerals, and phytochemicals frequently found in fiber-rich foods.

In addition, increase your fiber intake slowly to avoid digestive upset. Be sure to drink plenty of fluids to help move fiber along the digestive tract.

Fiber Tip: White or Whole Wheat?

The amount of fiber in your bread, cake, or muffin has a lot to do with how much fiber is in the flour used to make it. Whole wheat flour is basically the whole grain ground into fine particles. As a result, it contains the nutrients found in the whole grain.

White flour, on the other hand, has the bran and germ removed, to produce a finer flour that gives a more delicate texture to foods. In removing the bran and germ, much of the fiber and many of the nutrients are also removed. Most white flour, however, is enriched. That is, the thiamin, riboflavin, niacin, folic acid, and iron that is lost during refining is added back. But most of the fiber remains among the missing.

In the interest of combining good health with the pleasure we get from eating—which, after all, is important to staying with a healthy eating plan—all the grain foods you eat don't need to be whole grain. Try to choose whole grains for half of the recommended 6 to 11 servings of grain foods you eat each day. At a minimum, eat at least 3 servings of whole grain foods daily. Your favorite white-flour products can make up the rest of the grain foods you enjoy in your daily diet.

What Is Your Fiber Intake?

Use the chart on the following page to track your daily fiber intake. Once you have an idea of how much dietary fiber you already eat and where it comes from, making small dietary changes to get more fiber will be easy.

Food Group	# of Servings a day	Value	Approx. Fiber (g)
Whole Grains Serving size: 1 slice whole wheat bread, 1/2 cup cooked bulgur, brown rice or other whole grain; 1/2 bran or whole grain muffin	________	x 2.5 =	________
Refined Grains Serving size: 1 slice white bread, 1/2 cup cooked white or spinach pasta, white rice or other processed grains, and 1/2 bagel or muffin)	________	x 1.0 =	________
Breakfast Cereals Serving size: check food label for serving size and amount of fiber per serving	________	____ gm	________
Vegetables Serving size: 1/2 cup cooked vegetables; 1 cup raw vegetables	________	x 2.0 =	________
Fruits Serving size: 1 whole fruit; 1/2 grapefruit; 1/2 cup berries or cubed fruit; 1/4 cup dried fruit	________	x 2.5 =	________
Dried Beans, Lentils, Split Peas Serving size: 1/2 cup cooked beans, lentils, split peas	________	x 7.0 =	________
Nuts, Seeds Serving size: 1/4 cup nuts and seeds; 2 Tbsp. peanut butter	________	x 2.5 =	________
Total Fiber/day =			________

Source: Adapted from *Consumer Reports on Health Newsletter,* March 1995, Volume 7, Number 3

Fiber Tip: Choose Food First!

Why not just take a fiber supplement to make sure you meet your fiber needs? It's not that simple. All the health benefits linked to a fiber-rich diet aren't necessarily from the fiber.

Studies show a diet that contains plenty of fiber-rich foods such as whole grain breads and cereals, vegetables, and fruits may help reduce your risk of diseases such as heart disease

and some types of cancer. But it's not so clear th
the fiber in these foods at work. These foods also
abundance of other disease-fighting components su
tochemicals and antioxidant vitamins and minera
vitamins A, C, or E and selenium. Antioxidants
protect against diseases like heart disease and can
can squelch cell-damaging free radicals that are produced as part of normal bodily processes before the radicals have a chance to cause harm.

While your doctor may recommend fiber supplements for special health needs, use them as they are named—as supplements to a healthy diet with plenty of fiber-rich foods. For optimum health, choose food first!

Did you find that you need to increase your fiber intake? If so, consider these tips:

- Look for the words whole grain, whole grain oats, or whole grain wheat on the label of breakfast cereals.
- Look for bread labeled whole wheat or whole grain. Bread labeled wheat is not necessarily whole grain. Check the ingredient list on breads to make sure whole grain flour is the first ingredient used.
- Try brown rice versus white rice for a change.
- Choose the whole fruit instead of the juice for a lot more fiber.
- Eat the whole vegetable or fruit. The peel on your baked potato, cucumber, apple, or other vegetable or fruit is where a lot of the fiber is found.
- Get creative. Add legumes, vegetables, and fruits to casseroles, muffins, and breads. Cook up fruit sauces as toppings for chicken or desserts. Snack on a variety of dried fruits—prunes, apricots, and figs.
- Read the Nutrition Facts panel on the food label. Some food products, such as high-fiber breakfast cereals, are made specifically to help people get enough fiber. Foods that contain at least 2.5 grams of fiber per serving are a "good" source of fiber. Foods with 5 grams or more of fiber per serving are "high" in fiber.

Oat Bran—The "Whole" Story

Remember the oat bran craze of the 80s? Everything from beer to pretzels was made with oat bran, in an attempt to share in the glory of research that showed oat bran helped reduce blood cholesterol levels. The bottom dropped out of the oat bran market when one study suggested oat bran didn't have the desired effect. The study, however, was conducted on people with normal blood cholesterol, and oat bran really only helps reduce blood cholesterol in people with elevated levels. So while oat bran still offered promise for managing blood cholesterol levels, the public failed to understand the subtleties of the research and quickly grew disenchanted with oat bran.

Fast forward to 1997. That's when the Food and Drug Administration (FDA) breathed new life into oat bran and whole grain oats. Based on substantial research, the FDA approved the use of a new health claim on food labels that states foods rich in whole grain oats or oat bran, when eaten as part of a diet low in fat and cholesterol, may reduce risk of heart disease. But this time, there's an important caveat: Foods that wish to bear the health claim must contain at least 0.75 grams of beta-glucan per serving, a type of soluble fiber found in whole oats, or they can't make the claim.

So, goodbye oat bran beer. Hello oatmeal, Cheerios, Toasted Oatmeal, and some granola bars. These foods are among the few that currently qualify for the new claim. As soon as they are ready for market, you can expect more products such as low-fat oatmeal muffins and breads to bear the claim. To get the cholesterol-lowering benefits of whole grain oats and oat bran, you need at least 3 grams of beta-glucan fiber a day. The minimum of 0.75 grams per serving means you could need to eat as many as four servings a day of foods rich in whole grain oats or oat bran.

Chapter Three
Other Sweeteners

Sugar Alcohols: Not Sugars

When you look on the ingredient list of sugar-free candies, chewing gums, jams and jellies, you may see words like sorbitol, mannitol, and xylitol. They are ingredients used to sweeten a product...but they're not sugars. Along with maltitol syrup, lactitol, isomalt, and hydrogenated starch hydrolysates, they belong to a class of sweeteners called sugar alcohols. (They don't contain ethanol as alcoholic beverages do; the alcohol refers to their chemical structure.)

Sugar alcohols are found naturally in a wide variety of fruits and vegetables. They are also produced commercially from carbohydrates such as sucrose, glucose, and starch. Because sugar alcohols aren't absorbed as well as simple or complex carbohydrates, they contribute fewer calories to the diet. They also affect blood glucose levels less dramatically than other carbohydrates and therefore require little or no insulin for metabolism (see page 41). Thus, sugar alcohols are often used in foods for persons with diabetes.

Foods that contain sugar alcohols may be *sugar-free* but they aren't always low in calories. To be considered low calorie, foods must contain 40 calories or less per serving. Foods containing sugar alcohols may also contain other nutrients such as fat and complex carbohydrates which may contribute more than 40 calories per serving. Such foods must carry a statement on the food label advising consumers of this fact. Look for statements similar to "Not a reduced-calorie food."

Products containing sugar alcohols also may be labeled "Does not promote tooth decay." Sugar alcohols are not metabolized by the natural bacteria in our mouths that produce cavities.

When consumed in large amounts, however, sugar alcohols may produce gastrointestinal upset such as gas or diarrhea. The best advice: Enjoy foods sweetened with sugar alcohols in moderation.

Intense Sweeteners: Sweet Taste Without the Calories

Thanks to the work of food technologists, people today can freely enjoy sweet tastes, even when trying to cut calories. Although they taste very sweet, intense sweeteners—also called non-nutritive sweeteners, very-low-calorie sweeteners, and alternative sweeteners—contain no calories, or only a fraction of the calories provided by carbohydrates. Most intense sweeteners do not have the bulk that sugar does. Therefore, when cooking with intense sweeteners, it's best to use recipes that have been specifically developed for them.

While intense sweeteners can be part of healthy diet for most adults, infants and young children need the calories from carbohydrates for growth. Children over age two who eat a well-balanced diet may occasionally use foods and beverages sweetened with intense sweeteners.

Currently, there are three types of intense sweeteners approved for use in the U.S.:

Aspartame is a combination of two amino acids, phenylalanine and aspartic acid. Amino acids are the building blocks of protein. Phenylalanine and aspartic acid are found naturally in foods such as meat, skim milk, fruit, and vegetables. Although protein doesn't usually taste sweet, in aspartame these amino acids are joined in a way that gives them a sweet taste. During digestion, they are broken down and absorbed just like other amino acids.

Although aspartame contains the same amount of calories as carbohydrates (4 calories per gram), it is 180 to 200 times sweeter than sucrose. As a result, very little needs to be used to achieve the sweet taste desired. Therefore, aspertame adds almost no calories to foods. Aspartame breaks down under high temperatures, so it is primarily used in foods that aren't cooked, such as soft drinks, puddings, gelatins, frozen desserts, yogurts, hot cocoa mixes and more. It's also available as a tabletop sweetener, such as Equal. When cooking at home, add aspartame after a food is cooked to preserve the sweet taste.

People with phenylketonuria (PKU) must be cautious about consuming aspartame which contains phenylalanine. PKU is a rare metabolic disorder in which the body cannot metabolize phenylalanine. Foods containing aspartame must list aspartame in the ingredient list found on the food label and bear the warning: "Phenylketonurics: Contains Phenylalanine."

Saccharin is made from a substance that occurs naturally in grapes, and has been used as a sweetener for about 100 years. In 1977, the Food and Drug Administration posed a ban on saccharin because several studies suggested that saccharin in very large amounts may cause cancer in laboratory rats. Warnings about the findings were required to be featured on product labels. Human studies never confirmed the findings, and in 1991, the FDA formally withdrew the ban. The warning remains on product labels, though.

Saccharin is 300 times sweeter than sucrose. It contains no calories because it is not broken down by the body. Instead, it is excreted in the urine. Saccharin is used in soft drinks and is available as a tabletop sweetener, such as Sweet'n Low. Saccharin can be used in cooking and baking.

Acesulfame K was approved for use in the U.S. in 1988. It is 200 times sweeter than sucrose. Acesulfame K can be used by itself to sweeten foods. Also, it is often blended with other sweeteners. Like saccharin, acesulfame K is not broken down by the body and is excreted in the urine. Acesulfame K is used in candies, baked goods, desserts, soft drinks, and tabletop sweeteners, such as Sweet One. Acesulfame K is also heat stable and can be used in cooking and baking.

Chapter Four
On the Label

THE FOOD LABEL can be very helpful if you are trying to choose more fiber-rich and carbohydrate-rich foods. And since nutrition labeling is mandatory on almost every packaged food in the supermarket, you will not have a problem finding the nutrition information you need.

Most of the nutrition information appears on the Nutrition Facts panel found on the food label. The Nutrition Facts panel features the calories and the amount of specific nutrients in one serving of the food. The nutrients that appear include total carbohydrates, sugars, and dietary fiber.

The ingredient list on the food label provides the ingredients used to make the food product. The ingredients used appear in order of descending weight, from most to least. This information can be helpful in determining the type and to some extent, the amount of carbohydrates in a food product.

Additional nutrition information, such as nutrient content and health claims, may be found on the food label. A nutrient content claim helps describe a nutritional aspect of the food. Two examples are *sugar-free* and *high fiber*. These statements can help you choose between foods, since the terms mean the same thing on whatever food package they appear.

Health claims describe the potential health benefits of a food or nutrient, such as the reduction of some cancers with a fiber-rich diet. Health claims may only appear on qualifying food products.

First, the Facts

Nutrition Facts

The nutrition information in a serving of a food appears on the Nutrition Facts panel. The Nutrition Facts panel may list 25 or more nutrients. However, most food labels feature only nutrients that are required.

In terms of carbohydrates, total carbohydrate, dietary fiber, and sugars are the only carbohydrate figures that must appear on the food label. Others, such as soluble fiber, insoluble fiber, and sugar alcohols, may be listed or are mandatory if claims are made about these components on the food label.

% Daily Value

The Nutrition Facts panel provides not only the number of grams of total carbohydrates and dietary fiber in a serving of a food, but how it contributes nutritionally to a 2,000 calorie diet. That makes it easier to look at the big picture—the nutritional content of your *total* diet.

Nutrition Facts

Serving Size 1 cup (248g)
Servings Per Container 4

Amount Per Serving	
Calories 150	Calories from Fat 35
	% Daily Value*
Total Fat 4g	**6%**
Saturated Fat 2.5g	**12%**
Cholesterol 20mg	**7%**
Sodium 170mg	**7%**
Total Carbohydrate 17g	**6%**
Dietary Fiber 0g	**0%**
Sugars 17g	
Protein 13g	
Vitamin A 4% •	Vitamin C 6%
Calcium 40% •	Iron 0%

* Percent Daily Values are based on a 2,000 calorie diet. Your daily values may be higher or lower depending on your calorie needs:

	Calories:	2,000	2,500
Total Fat	Less than	65g	80g
Sat Fat	Less than	20g	25g
Cholesterol	Less than	300mg	300mg
Sodium	Less than	2,400mg	2,400mg
Total Carbohydrate		300g	375g
Dietary Fiber		25g	30g

Calories per gram:
Fat 9 • Carbohydrate 4 • Protein 4

The % Daily Value (%DV) column on the Nutrition Facts panel is the means by which you can do that. It tells you if a food has a lot or a little of a specific nutrient.

A 2,000 calorie diet is the reference for calculating the percentage of the DV a food supplies. Two thousand calories is used because this is a typical calorie level needed by women and sedentary men. Some larger food packages may also list in a footnote the % DV for several food components, including total carbohydrate and fiber, for a 2,500 calorie diet.

If a serving of a food contains 10% of the DV for total carbohydrate, it supplies 10% of the total amount of carbohydrates needed daily for good health. If it supplies more than 50% of the DV for fiber, as some high-fiber breakfast cereals do, you may meet more than half your daily fiber requirement just by eating one serving of the cereal!

DVs have been established for total carbohydrate and dietary fiber, but not for sugars. Since there is no recommended daily intake for sugars, the Nutrition Facts panel cannot list a % DV. Sugars are listed on the Nutrition Facts panel to help you consume sugars in moderation.

Nutrition and health experts agree that 60 percent of your total calorie intake should come from carbohydrates. That translates to a DV of 300 grams of carbohydrates in a diet that contains 2,000 calories a day. The DV for fiber is 25 grams a day. If your calories needs are higher or lower than 2,000 calories, you can precisely calculate how much total carbohydrate and fiber you need. Just multiply your calorie needs by 60 percent, then divide by 4 (the number of calories per gram of carbohydrate). For example:

1800 calories x 60% = 1080 calories ÷ 4 calories/gram
= 270 grams carbohydrate

Fiber needs are based on 11.5 grams per 1,000 calories:

1800 calories ÷ 1,000 calories = 1.8 x 11.5 grams = 21 grams fiber

Sizing Up a Serving

The serving size listed on the Nutrition Facts panel for a food is based on how much people actually eat of that type of food—not necessarily the amount recommended or the amount you eat. It's important because nutrition information applies to one serving. So, if a serving is one cup and you eat two, you will consume twice the amount of calories and other nutrients provided on the Nutrition Facts panel.

Label Lingo

The Ingredient List

The Nutrition Facts panel tells you the amount of sugars in a food. But the ingredient list tells you what type of sugars they are. Here's a tip on recognizing some of the more technical-sounding names: All ingredients ending in *-ose* are a type of sugar. How many of these sugars do you recognize?

- Brown sugar
- Confectioner's sugar
- Corn syrup
- Dextrin
- Dextrose
- Fructose
- Fruit juice concentrate
- Glucose
- Honey
- Lactose
- Turbinado sugar
- Invert sugar
- Raw sugar
- Cane sugar
- Corn sweeteners
- Crystallized cane sugar or juice
- Evaporated cane juice
- High-fructose corn syrup (HFCS)
- Malt
- Maltose
- Maple syrup
- Molasses
- Sucrose

Now when you read the ingredient list the information provided may be more useful.

Chocolate Chip Cookies

Ingredients: Enriched wheat flour (contains niacin, reduced iron, thiamine mononitrate [vitamin B1], riboflavin [vitamin B2]), sweet chocolate chips (made with sugar, chocolate, cocoa butter, dextrose, soy lecithin—an emulsifier), sugar, vegetable shortening, brown sugar, cornstarch, high fructose corn syrup, leavening, salt, whey, dextrin, soy lecithin, natural and artificial flavor, artificial color (contains yellow 5, Yellow6, Red 40, Blue 1), sodium benzoate and potassium sorbate (to preserve freshness).)

Nutrient Content Claims

Many food packages also carry nutrient descriptions such as "high fiber," "fat free," or "low calorie." These nutrient content claims, as they are officially called, describe the level of a nutrient in a serving of food. The claims can only be used if the food on which they appear meet strict definitions set by the FDA or the United States Department of Agriculture.

As far as carbohydrates are concerned, nutrient content claims are only allowed for sugars or fiber in a food.

What Does it Mean: Carbohydrate Claims on the Food Label

Label Term *Sugar free, free of sugar, no sugar, zero sugar, without sugar, sugarless, trivial source of sugar, negligible source of sugar, dietarily insignificant source of sugar*
What It Means Contains less than 0.5 grams sugar per serving

Label Term *No added sugar, without added sugar, no sugar added*
What It Means ➤ No sugars added during processing, including ingredients that contain sugars such as fruit juice. ➤ Processing does not increase the sugar content above the amount naturally present in the ingredients. ➤ The food that it resembles and for which it substitutes normally contains added sugars.

Label Term *Reduced sugar, reduced in sugar, sugar reduced, less sugar, lower sugar, lower in sugar*
What It Means Contains at least 25 percent less sugar per serving than the food to which it is compared. ➤ Main dish products and meal products containing at least 25 percent less sugar per 100 grams.

Label Term *High fiber, rich in fiber, excellent source of fiber*
What It Means Contains 5 grams or more per serving.

Label Term *Good source of fiber, contains fiber, provides fiber*
What It Means Contains 2.5 grams to 4.5 grams per serving.

Label Term *More or Added Fiber*
What It Means Contains at least 2.5 grams more fiber per serving than the food to which it is compared.

You may see the term "light" on some foods. "Light" carries an official definition that can be used in connection with calories, fat, or sodium. But, it can continue to be used to refer to color or texture in foods that have a long history of use of the term. For example, a package of brown sugar may bear the descriptor "light brown sugar," which refers only to its color, not its calorie content.

Health Claims

Along with the Nutrition Facts panel and nutrient content claims, some food packages may also feature specific phrases that explain the health benefits of eating the foods. The phrases are actually health benefit claims that the FDA has approved to help educate consumers about certain relationships between what you eat and chronic diseases.

Health claims first became popular in the 1980s, when a breakfast cereal company linked up with the National Cancer Institute to advertise the potential of a low-fat diet rich in fiber-containing foods to reduce the risk of cancer. The advertised cereal enjoyed a significant sales spurt following the campaign, which prompted other manufacturers to follow suit.

Subsequently, the FDA acted to control the types of claims that could appear on food, to ensure a strong scientific base for health claims. To date, ten written health claims have been approved for use on food packages.

Three of the health claims are specific to carbohydrate-rich foods. The information they provide may not be news to many nutrition-savvy people. But the claims reach a very wide audience through the food label. Registered dietitians, therefore, have high hopes that health claims will help guide Americans in choosing foods for good health.

Carbohydrate-Rich Foods & Health: Look to the Label

Claim Low-fat diets rich in fiber-containing grain products, fruits, and vegetables may reduce the risk of some types of cancer.
Typical Foods that May Include the Label Claim Whole grain breads and cereals; Fruits; Vegetables

Claim Diets low in saturated fat and cholesterol and rich in fruits, vegetables, and grain products that contain some types of dietary fiber, particularly soluble fiber, may reduce the risk of heart disease.
Typical Foods that May Include the Label Claim Fruits; Vegetables; Whole grain breads and cereals

Claim Soluble fiber from foods such as whole grain oats, as part of a diet low in saturated fat and cholesterol, may reduce the risk of heart disease.
Typical Foods that May Include the Label Claim Oatmeal; Oat bran; Breakfast cereals made from whole grain oat flour

Chapter Five

Carbohydrates

Fueling Up for a Healthy Weight & Physical Fitness

AFTER READING THE preceding chapters, you know plenty about carbohydrates—what they are, where they come from, some of their health benefits and how to find carbohydrate-rich foods. But chances are, you still have plenty of questions.

In particular, what about the latest high-protein, low-carbohydrate diets touted for weight control and athletic performance? Are they a good idea? Whatever happened to carbohydrate-loading?

This chapter looks at the impact of carbohydrates on weight and physical performance.

High-Protein Diets—They're Baaaack!

Anyone who is interested in weight management these days—whether for personal or professional reasons—probably hasn't missed the return of the high-protein diet. Today's popular books that advise limiting carbohydrates and eating plenty of protein-rich foods to help manage weight (as well as a variety of illnesses) sound disturbingly familiar. While today's versions include pseudo-scientific facts to support their claims, the real truth is they are no different from the high-protein weight-loss fads of the 1970s.

The premise of a high-protein, low-carbohydrate diet is that a high-carbohydrate intake leads to increased insulin levels, particularly in people who are insulin resistant (see "About Insulin Resistance," page 31). The increased insulin, in turn, supposedly promotes the conversion of excess carbohydrate in the diet to fat, which is then stored as body fat, thereby contributing to obesity.

This theory, however, falls short of the science. The fact is that your body uses most of the carbohydrates that you eat for energy and bodily processes. Even if excess calories are eaten, carbohydrates are burned in preference to fat in the diet. And dietary fat, which is found in many high-protein foods, is stored more easily as fat in the body than excess carbohydrates. Therefore, eating a low-fat, high-carbohydrate diet can help with weight maintenance—not a low-carbohydrate diet.

Several of these popular diets promote very-low-carbohydrate diets not only as a way to avoid gaining weight but also to lose weight. Such diets do appear to promote quick weight loss because you lose body water when you don't eat enough carbohydrates. But it's not a permanent weight loss: The water—and weight—quickly returns once you return to a balanced diet.

Loss of body fat can occur on low-carbohydrate diets over time. But the reason for the weight loss doesn't have anything to do with carbohydrates per se. It's that the diets usually provide fewer calories than are needed to maintain weight. There is nothing about a low-carbohydrate diet that is particularly advantageous when you want to cut calories. Calorie control can be achieved on a balanced diet. That means weight loss can be achieved in a much healthier way, both psychologically and physically.

In psychological terms, one primary danger of low-carbohydrate diets is that once again dieters are led astray in the search for permanent weight management. People experience a great deal of difficulty staying on diets that require them to severely restrict their food choices, especially when the restrictions include favorite foods. Diet "failure" is discouraging. It can even cause people to permanently give up at attempts to manage their weight.

Further, history shows us that when dieters are severely restricted, they tend to overcompensate once they begin eating "forbidden" foods again. This may lead to a binge cycle that creates more problems with managing weight, and ultimately health.

When people follow a low-carbohydrate diet for an indefinite time, they may also increase their fat intake. That's true even if the advice cautions dieters to choose only lean sources of protein. Taste bud fatigue can take over, driving low-carbohydrate aficionados to seek more tasty (translation: higher-fat) sources of protein. A substantial body of evidence shows that people tend to eat more calories on a high-fat, low-carbohydrate diet. What's more, a high-fat diet is linked with heart disease and some types of cancer.

About Insulin Resistance

Insulin is a hormone that allows glucose to travel from the blood into the cells, where the glucose is converted into energy or stored as glycogen. In people who are insulin resistant, their bodies don't allow the insulin to do its job. To compensate, additional insulin is produced, driving up blood levels of insulin.

Diet book authors claim high levels of insulin promote the storage of glucose as fat. The facts argue otherwise. First, you must eat more calories than you need to make and store body fat. Second, carbohydrates get used before fat does. So if you eat too much, it's the fat in your diet that is getting stored as body fat first, not the carbohydrates. The bottom line is that it's the amount of calories, rather than the source, that matters.

Other In-the-News Myths about Low-Carbohydrate Diets

Today's popular high-protein, low-carbohydrate diet books feature a variety of claims about the potential benefits of their prescribed regimens. Just to set the record straight, here's the scoop on some of the claims.

Claim *Refined carbohydrates, such as those in white flour breads, bagels, muffins, and snack foods, can cause rapid weight gain.* **Fact** Weight gain happens when intake is greater than output, regardless of food type. Anyone can gain weight from eating any type of food, provided they eat more than the energy they expend. There is no scientific evidence that carbohydrates stimulate appetite and lead to more fat storage and weight gain.

Claim *You can "reset your genetic code" with a low-carbohydrate, high-protein diet.*
Fact Your genetic code is impossible to reset. Your genes are your genes, and they haven't changed for thousands of years.

Claim *"You must treat food as if it were a drug. You must eat food in a controlled fashion in the proper proportions—as if it were an intravenous drip."*
Fact In the high-paced, stressed-out lifestyle of the '90s, the last thing Americans need to follow is a strict, controlled regimen, especially when eating can and should be a pleasurable experience.

Claim *Many people have gained weight on a low-fat diet.*
Fact Successful low-fat diets for weight control need to limit calories, too. If dieters increase their total calories, no matter if it's from carbohydrate, protein, or fat, they will gain weight.

Claim *High-protein diets have worked for hundreds of people.*
Fact Scientifically validated studies are the cornerstones for basic nutrition principles, not anecdotes from friends, family, and colleagues.

Claim *One of the greatest weight-loss benefits of a high-protein diet is that it is "brain energizing."*
Fact Glucose is the only form of energy the brain can use; its primary source in the diet is carbohydrates, not protein.

Claim *The insulin released as a result of eating carbohydrates causes the body to make "bad" eicosanoids, a substance involved in a wide range of conditions, including heart disease, cancer, arthritis, lupus, multiple sclerosis, eczema, alcoholic cravings, dull hair, dry skin, and brittle nails.*
Fact Eicosanoids are just one part of a complex system. The body is not generic—one thing cannot affect everything else. What's more, there are no studies to suggest eicosanoids are dangerous in any way.

The Truth About Carbohydrates & Weight

Foods rich in complex carbohydrates play an important role in weight management. These foods are frequently low in calories and fat. What's more, they often contain significant amounts of fiber. Fiber-rich foods may help you feel more satisfied and thereby help discourage overeating.

Complex carbohydrate foods also tend to be rich in essential nutrients, another important consideration for calorie watchers. Nutrient-dense diets are key to good health. Fad diets that restrict intake of complex carbohydrates may also restrict intake of other key nutrients.

On the other hand, foods rich in simple carbohydrates like candy and soda do not contain a significant amount of important nutrients other than calories. Therefore, they should be eaten in moderation. Is there any truth to the claim that foods containing primarily simple carbohydrates play a unique role in promoting overeating and obesity? Fortunately, no. To explain, let's look at some of the research in detail.

Does Sugar Make You Fat?

One way scientists have explored the impact of sugars and sweet tastes on the overall diet is to give "pre-loads" of sugar-sweetened drinks before a meal and then examine how many calories are consumed and what foods are chosen during the meal. Studies indicate that no matter what the form of the pre-load carbohydrate—whether it is glucose, fructose, sucrose, or starch—food intake following consumption of the "pre-load" carbohydrate drops according to how many calories were consumed.

For example, studies show that when adults consume soft drinks 20 to 60 minutes before a meal, their caloric intake at the meal decreases. The implication is that if you drink a soda before a meal or snack, you will likely eat less at the meal or snack to compensate for the calories consumed from the soda.

Children in particular appear to be very good at this type of compensation. But they take it a step further. They not only tend to eat fewer calories after they eat highly-sweet foods, but also choose less sweet foods.

Other experimental studies show that obese people are no different from normal weight people in their preference for, or ability to detect, sweet tastes. Instead, they seem to like fat more than normal weight individuals. Obese women prefer sweet/fat combinations such as cakes and other desserts; obese men prefer high-protein, fatty meats.

Surveys and population studies also suggest it is the fat in the diets of obese people, not the carbohydrate, that may promote the consumption of too many calories. Further, people who eat more carbohydrates appear to be leaner.

The Bottom Line

To lose weight or maintain weight, a high-carbohydrate, low-fat diet along with regular physical activity ranks as the best choice. According to national dietary guidelines, a high-carbohydrate, low-fat diet is also recommended to reduce risk of chronic disease, such as heart disease, and promote overall good health.

With that said, if too many calories are consumed—whether from protein, carbohydrate, or fat—weight gain can result. To help control your total caloric intake, watch your portion sizes. Also, try eating plenty of fruits and vegetables. Besides having numerous nutritional benefits, they are low in calories, mostly fat free, and can help fill you up.

Getting Started

These tips can help you put the advice for weight management and weight loss into a lifestyle that works for you:

Assess your current habits. Start by keeping track of what, when and why you eat or drink. Do you snack when you're bored? Do you lead a busy life that dictates quick meals that are often high in fat? Assess your physical activity. Do you take a daily walk, or are you more likely to find yourself in front of the television when you're at home?

Set reasonable goals. Based on your current lifestyle, determine what to change. You don't have to give up current habits, you may just need to modify them to introduce a bit more moderation and balance. For example, if you love hot fudge sundaes, you can continue to enjoy them—but probably not every night!

Move slowly. That doesn't refer to physical activity. It means to make gradual changes in your habits. Small steps are easier to take than giant steps. The easier it is, the more likely you'll do it.

Monitor your progress. Give yourself a star on your calendar for every day you follow your plan for healthy eating and physical activity. If you get off track, don't worry. Just start back up again. When you see you've got a small step well-established as part of your lifestyle, take the next one. Over time, that's how you'll get where you want to go.

Reward yourself. Recognize your hard work with special treats: A relaxing bath at the end of a day, a CD at the end of a week, a new outfit at the end of a month—whatever appeals to you! But remember, feeling great is the best reward of all. You'll see your spirits and energy levels soar after adopting a healthy lifestyle.

Seek help from a professional. If you're having trouble eating healthfully, see a registered dietitian. Likewise, a certified exercise professional may be able to help you get started and stay with a program for physical activity.

Evaluating Diet Fads

Like much nutrition nonsense that has come before, the high-protein diet fad promises more than it can deliver. Indeed, in the search for a healthy weight, people have seen many diet fads come and go over the years. The next time a new diet comes your way, consider these points before joining in.

- Do the claims seem reasonable? If claims for improved performance and weight loss sound too good to be true, they probably are too good to be true.
- Does the diet promise quick weight loss? Sound weight loss programs aim for losing no more than one-half to one pound per week.
- Does the diet plan recommend physical activity? Exercise is the best predictor of success in maintaining weight loss.

- ➤ Are only certain foods emphasized? Are some foods not allowed? Eating foods you enjoy, in moderation, makes it easier to stick with a diet. Not allowing certain foods may increase the likelihood that you will "cheat," binge, or even give up on the diet.
- ➤ Can you sustain the regimen for the rest of your life? Weight loss or maintenance depends upon sticking with an eating and training plan over the long term.

Fueling Physical Activity

The current high-protein, low-carbohydrate diet fad isn't aimed only at people interested in weight control. It is also being widely touted as *the* way to achieve peak athletic performance.

One popular approach features "40/30/30" diets—diets with 40 percent of calories from carbohydrate, 30 percent from protein and 30 percent from fat. One such diet book author claimed a 40/30/30 regimen led to a winning streak by the women's swim team at Stanford University, after years of losing to the University of Texas. The claim, by the way, was responsible for the tremendous amount of attention the book initially received. The author failed to note that before the winning streak, the Texas coach and several athletes of national caliber joined the Stanford team.

Similarly, many of the other claims about the 40/30/30 approach lack a certain grounding in the truth. The real truth is that for athletes who are already meeting their nutritional needs, real changes in performance only come through training.

When it comes to eating for athletic performance, nothing beats a diet that contains plenty of grain products, fruits, and vegetables. As you have already learned, the carbohydrates in these foods serve as the first and best source of energy for the body. They are stored as glycogen (stored carbohydrate), the primary source of energy for working muscles during intense physical activities. And they help replenish the glycogen following intense activities, to help ensure continued optimum performance.

What type of athlete are you? If you're a "30-minute-a-day" type (the minimum amount of activity recommended to promote health), you will of course have different nutrition concerns than a highly-trained competitive athlete. The following serves as a simple guide. If you have questions, talk to a registered dietitian who is knowledgeable in sports nutrition.

For Less-to-Moderately Active, Healthy People If you're like many Americans who are getting no more than the minimum recommended 30 minutes a day of physical activity, you are probably more concerned about managing your weight through exercise than about achieving peak performance. You can continue to meet your nutritional needs by following a diet as illustrated by the Food Guide Pyramid. Your intake will be about 55 to 60 percent carbohydrate, 30 percent fat, and 10 to 15 percent protein.

For Recreational Athletes If you enjoy two to four hours of moderate to vigorous physical activity a week, you will likely need more calories than the average person. You should continue to eat the same percentage of carbohydrate, fat and protein but may need to increase how much you eat to maintain your weight. Be sure to drink plenty of fluids. Experts recommend drinking about two cups (16 ounces) of fluids two hours before exercise, and as much as possible during and after exercise to replace fluid lost through sweat. Even when they don't feel thirsty, athletes should make an effort to drink plenty of fluids. Fluid loss can range from one to four pounds per hour; for every pound lost, two cups of fluid is needed for rehydration.

For Competitive Athletes College athletes, marathoners, cyclists, and others generally require significantly more calories (as much as 3,500 to 4,000 calories a day). Studies also show that diets containing 70 percent of calories from carbohydrates improve performance compared to a diet containing between 40 and 45 percent of calories from carbohydrates. During intensive exercise lasting more than one hour, it may be advisable to consume 30 to 60 grams of carbohydrates each hour to delay fatigue. Sports drinks are often considered ideal for this purpose since they simultaneously supply carbohydrates and fluid. Foods with a high glycemic index (GI) may also be appropriate to maintain blood glucose levels.

Carbohydrate "re-loading" after an event is equally important to replenish glycogen stores. For maximum glycogen resynthesis, research suggests that foods with a high to medium glycemic index should make up the majority of the post-exercise diet. Adequate fluid intake remains essential. Refer to page 43 for more information on the glycemic index of foods.

Carbohydrate Loading for Optimum Performance

Carbohydrate loading is a way to give your muscles extra fuel when preparing for an endurance event. For an endurance activity that lasts more than 90 minutes, increasing your daily intake of complex carbohydrates during the week prior to competition will help prevent glycogen depletion and allow you to compete at your best. Carbohydrate loading consists of a daily diet that is high in carbohydrates, low in fat and balanced with adequate protein. It is also necessary to gradually decrease the amount of training the week before an event.

Get about 70 percent of your calories from carbohydrates, drink extra fluids to hydrate your body, and make sure to get enough rest. If you need help in planning a diet for carbohydrate loading, consult a registered dietitian experienced in sports nutrition.

Chapter Six

Reducing the Risk for Disease

What Role Do Carbohydrates Play?

HEART DISEASE, CANCER, diabetes, diverticulosis, hemorrhoids, hypoglycemia. What do all these illnesses have in common? A high-carbohydrate diet helps to reduce the risk for them all.

Heart Disease—Still Number One After All These Years

It may be number one, but in this case, that's not good news. Even after years of study and nationwide campaigns to stop heart disease—the number one killer of Americans—it still reigns supreme. About one fourth of the 158 million people in this country have some form of cardiovascular disease. It accounts for about 950,000, or close to one-half, of all deaths annually in the U.S.

There are a variety of reasons people develop heart disease. Some of them you can control, some you can't. Genetics—or the tendency towards the disease that you inherit from your parents, being a black American, and getting older are risk factors that are out of your hands. But you can control whether you smoke cigarettes, lead a sendentary lifestyle, are overweight, drink too much alcohol, and eat a diet high in total fat and saturated fat, all of which have a significant impact on the risk for heart disease. Likewise, if you have high blood pressure, high blood cholesterol (over 200 milligrams per deciliter), or diabetes, you can take steps to control those problems that also contribute to heart disease risk.

On the diet scene, reducing saturated fat and total fat intake has been the primary aim of heart health experts. That's because reducing the saturated fat in your diet can help lower blood cholesterol. And reducing your total fat can help you consume fewer calories and therefore help with weight control as well as limit your saturated fat intake. Dietary cholesterol may raise blood cholesterol levels in some people, although generally to a lesser extent than saturated fat.

Too much cholesterol in the blood increases risk for heart disease because it can collect on the walls of your arteries and other blood vessels. As it builds up, fatty plaques form that can eventually block the flow of blood through the vessel. When vessels become about 75 percent blocked, chest pains may occur. Or you may have no warning at all. If a clot stops up a narrowed artery, blood can't get to the heart. A heart attack then occurs. If a clot happens on the way to the brain, a stroke results.

So where do carbohydrates come in? Like the diet recommended in the Food Guide Pyramid, a heart healthy diet is based on foods rich in carbohydrates and that contain plenty of fiber. And it's a heart-healthy diet is low in saturated fat and total fat.

In addition, carbohydrate-rich foods—specifically, those that contain plenty of fiber—often come packed with other substances important in the fight against heart disease. Antioxidants such as vitamin E and selenium, found in whole grains, and beta-carotene, found in fruits and vegetables, may help control cell-damaging free radicals. Free radical damage may play a key role in starting the build-up of plaque on vessel walls. Folic acid, a vitamin found in abundance in legumes, green leafy vegetables, fortified breakfast cereals and enriched grain products, is being studied for its ability to reduce homocysteine in the blood, which appears to be linked to heart disease. Other phytochemicals in fiber-rich foods are also being scrutinized for their potential to help protect against heart disease as well as other chronic diseases.

Then there's soluble fiber. Soluble fiber found in dried beans and peas, oats, barley, and many fruits and vegetables may boost the heart-disease-fighting impact of a diet low in saturated fat and cholesterol.

And don't forget whole grain oat products. As part of a diet low in saturated fat and cholesterol, foods made from whole grain oats or oat bran have a special ability to reduce blood cholesterol levels.

Managing Diabetes—The Carbohydrate Connection

Just a few years ago, people with diabetes got some good news. Diabetes experts agreed that sugar (sucrose) could be substituted for other carbohydrates as part of a healthy eating plan for people with diabetes. This represented a major turnaround in thinking. Sugar was generally thought to be a forbidden food for people with diabetes. Many people even mistakenly believed that sugar caused diabetes.

The news isn't license to eat sugar freely, however. It just put it on par with other types of carbohydrates. Sugar does cause blood glucose (blood sugar) levels to rise, but so do all other carbohydrates, although the rate at which blood glucose rises after eating them differs.

To understand this, let's back up a little. Diabetes is a condition in which the body either does not produce enough insulin to move glucose from the blood into body cells, or the body cannot efficiently use the insulin it does make. As a result, glucose builds up in the blood. We do not know why diabetes occurs, but we do know that eating sugar doesn't cause it. We also know that if it goes untreated, high blood glucose levels can cause a multitude of health problems. Heart disease, eye problems (even blindness), circulatory problems, foot problems, and kidney disease are among the many illnesses that may occur.

There are three types of diabetes:

Type 1 diabetes occurs when the body can't make insulin, or at least not enough. This type often begins in childhood or early adult years, although you can get it at any age. Insulin shots are needed daily.

Type 2 diabetes develops slowly and usually surfaces after the age of 40. Being overweight is a common risk factor for this type of diabetes. It usually can be controlled by diet, losing weight, and exercising. Sometimes, however, insulin or other medications are necessary.

Gestational diabetes occurs during some pregnancies, usually due to changes in hormone levels. Although it usually disappears after the baby is born, it needs to be carefully controlled during pregnancy. Women with gestational diabetes often develop diabetes later in life, and usually in later pregnancies.

No matter what the type of diabetes, an eating plan to manage the condition is a must. Generally speaking, people with diabetes should follow an eating plan that is high in complex carbohydrates and low in fat. This is especially true for people who are not insulin dependent. For some people with diabetes, the proportions of carbohydrate, protein, and fat may need to be more individualized. A registered dietitian can help design an eating plan that meets the specific needs of a person with diabetes.

Now let's return to the subject of carbohydrates. The new freedom to allow sugar in the diets of people with diabetes came out of research on the glycemic index (GI) of foods. This research showed that the way carbohydrate foods affect blood glucose cannot be predicted according to whether they are simple or complex carbohydrates.

The GI of a food tells you its relative ability to raise blood glucose levels. The rate is compared to the rate for white bread, which is arbitrarily set at 100. If the GI is higher than 100, it means the glucose is released more rapidly, and that the carbohydrate in the food raises blood glucose levels faster. If it is lower than 100, the glucose is released more slowly, and the carbohydrate in the food raises blood glucose more slowly. Sucrose actually has a lower GI than mashed potatoes!

In actual practice, the concept doesn't work so simply. The rise in blood glucose is also affected by meal size, other foods in the meal, amount of fat in the meal, and individual health status. Still, the concept is valid, and may be used as a tool by a registered dietitian when designing an eating plan for people with diabetes who have trouble controlling their blood sugar, as well as for some competitive athletes. Check with your doctor or dietitian if you have questions about the glycemic index.

Glycemic Index of Common Foods

Food	Glycemic Index
Glucose	138
Corn Flakes	119
Potato (mashed)	100
Bread, white	100
Sugar (sucrose)	92
Oatmeal	87
Ice Cream	87
Rice, white	81
Banana	76
Peas, green	68
Orange	62
Spaghetti, white	59
Apple	52
Kidney beans	42

Source: *Carbohydrate News*, Issue Two. Canadian Sugar Institute, 1996.

Hypoglycemia: Do You Have It?

What is hypoglycemia? Do you have it?

If you ask these questions randomly to people you meet, you'll likely find hypoglycemia is a common disease, at the root of anxiety, headaches and chronic fatigue. They will also tell you it is caused by eating foods with sugar.

Wrong on all counts. First, hypoglycemia isn't a disease, it's a condition. Between meals, blood sugar levels naturally drop. But they remain fairly constant between 60 and 100 milligrams per deciliter (mg/dl). If levels drop below 40 mg/dl, hypoglycemia does occur, signaled by symptoms such as sweating, rapid heartbeat, trembling and hunger. These symptoms happen because there is not enough glucose available for cells to produce energy.

People with diabetes may experience hypoglycemia as a result of taking too much insulin, exercising too much, or by not eating enough. In other people, hypoglycemia is linked to problems such as liver disease or pancreatic cancer.

A condition called reactive hypoglycemia also occurs, although rarely. In this disorder, the body oversecretes insulin after a large meal. Blood sugar then drops well below normal. Shakiness, sweating, rapid heartbeat, and trembling may occur about two to four hours after eating. Don't confuse these symptoms with extreme hunger, characterized by stomach rumbling, headache, and feelings of weakness, which usually occurs from six to eight hours after a meal.

In all, true hypoglycemia is very rare unless you have diabetes. If you think it's a problem for you, talk to your doctor. Be wary of clinics that "diagnose" you with "sugar-induced" hypoglycemia or attribute emotional or other problems to the "sugar blues."

Problems Down Under: Diverticulosis, Hemorrhoids & Colorectal Cancer

For "problems down under," the benefits of carbohydrates can be summed up in a single word: Fiber. When recommended amounts of fiber are eaten, the insoluble portion works to help prevent diverticulosis, hemorrhoids, and colorectal cancer.

Diverticulosis

Insoluble fiber helps prevent constipation. Since fiber is not broken down by digestive enzymes and absorbed into the body, it adds bulk to the contents of the intestine. In addition, the insoluble fiber draws water into the intestine, helping to keep the contents soft. The result: A larger, softer stool that moves through the intestinal tract more quickly and more frequently.

As such a stool moves through the large intestine, it exerts less pressure on intestinal walls than a hard, compact stool. Diverticulosis—tiny sacs in the intestinal wall—is one consequence of too much pressure in this area. The tiny sacs can become infected and quite painful. When this happens, your doctor may advise a special diet that is lower in fiber until the flare-up subsides. A registered dietitian can help if this is a problem for you.

Hemorrhoids

Hemorrhoids—a painful swelling of veins near the anus—can also occur when a person must strain to eliminate hard, compact stools. Again, fiber plays a major role by creating large, soft stools that move through and out of the intestinal tract easily, without the need to strain.

Colorectal Cancer

Finally, eating plenty of fiber may reduce your risk for cancer of the colon and rectum. This benefit stems from the same actions described above. Because fiber helps food wastes move through the intestinal tract more quickly, there is less time for potentially harmful substances to come in contact with the intestinal walls. What's more, bulkier stools help dilute the concentration of these substances.

The potential anti-cancer benefits of a high-fiber diet may also be linked to the fact that a high-fiber diet is usually low in fat. Nutrition experts advise a low-fat, high-fiber diet as your best dietary defense against cancers "down under."

For Healthy Chompers

When you think of cavities, what's the one food that comes to mind? Probably candy. For years, we have been warned to go easy on candy and sugary snacks if we want strong, healthy teeth. But there's much more to that story.

First, the good news. Over the past 20 years, Americans on the whole have had fewer cavities. Still, the cost of taking care of our dental health ranks second only to the costs of cardiovascular disease in this country.

To prevent cavities, advice in the past focused on brushing our teeth twice a day, flossing and, as discussed above, limiting the amount of sucrose eaten. But in recent years studies have shown that the decrease in cavities seen in the U.S. and other developed countries is due to a large extent to an increased exposure to fluoride from water, toothpaste and mouthwashes. Fluoride strengthens the teeth, making it more resistant to the acid produced by mouth bacteria.

Even so, fluoride doesn't remove all threat to the health of our teeth. We still have to consider what we eat and how we eat it.

A balanced diet that includes other important nutrients such as vitamins A, C, and D and calcium is critical to tooth and gum health throughout life.

When it comes to carbohydates, the key consideration concerns how long they are in the mouth, not what type of carbohydrates they are. Both simple and complex carbohydrates, not just the simple sugar sucrose, serve as food for mouth bacteria. When bacteria act on carbohydates, they produce the acid that "eats" through tooth enamel to produce cavities.

Considering this, it would appear that advice to eat a high-carbohydrate diet would be contrary to dental health. The point we need to understand, again, is that how long carbohydrates stay in the mouth is the key.

That brings us to two important pieces of advice about controlling the effect of carbohydrates on dental health: First, consider the stickiness of a food. Foods like caramels and raisins stick to the teeth and thereby can remain in the mouth longer than foods like soft drinks. That doesn't mean you can't continue to enjoy sticky, carbohydrate-rich foods on occasion. Just brush your teeth with a fluoride toothpaste after eating them. If that's not possible, rinse your mouth with water. Chewing sugar-free gum may also help neutralize the acid produced by mouth bacteria.

Second, consider how frequently you eat carbohydrate-rich foods. Even if they don't stick to the teeth, constantly sipping on sugar-sweetened beverages (including fruit juices) means carbohydrates are constantly in your mouth. The best example of this is the tooth decay seen in infants who are put to bed with bottles containing carbohydrate-rich liquids, such as milk or juice. Likewise, frequent snacking on starchy or sugary foods can have a similar effect.

Finally, it's worth repeating that there's no substitute for dental care. Regular flossing and brushing with fluoride toothpaste along with regular visits to the dentist can pay off big.

Does Sugar Make Kids Hyperactive?

In the nutrition field, anecdotal reports—or what individuals *think* happen when they eat specific foods—often get reported as fact. The real fact is our feelings and reactions at any point in time are influenced by many things. Only well-controlled scientific studies can filter out the influence of these many factors to find the real impact of a food on our behavior.

This happened several decades ago when reports that sugar (sucrose) caused hyperactivity in kids made the headlines. The reports weren't based on scientific studies, just the opinion of some parents. Tightly-controlled research subsequently failed to show that kids who consistently ate high levels of sugar were hyperactive. Nor did hyperactivity occur after kids consumed single high doses of sugar.

One of the best studies in this area showed sugar may have an opposite effect on children. In line with other studies that show carbohydrates may have a calming effect, the study showed sugars tend to calm both children and adults. The effect could go unnoticed, however, due to other influences. For instance, the excitement of a birthday party or Halloween could override the calming effect of sugars.

Food & Mood—Do Carbohydrates Count?

Carbohydrates and Stress

Several years ago, an author got a lot of attention with a book that explained in detail how to manage your mood through food. The premise: eating high-protein, low-carbohydrate meals increase alertness and energy, whereas high-carbohydrate meals do the opposite.

Most experts aren't willing to go that far. The fact is, they are still not sure whether or how food may affect your mood.

That said, there is some preliminary research that suggests carbohydrates may make some people feel better during stressful times. But there is also research that shows no effect from carbohydrates. In other words, it's inconsistent.

Studies that have examined links between food and stress have shown that stress stimulates the breakdown of serotonin, a chemical that the body uses to soothe stress. Increasing your intake of carbohydrates may help replenish serotonin stores, thereby having a calming effect. But to correctly interpret what the results of all these studies mean, much more research needs to be conducted.

Carbohydrate Cravings

One area that has gotten a lot of attention from both scientists and the public is food cravings. People often report that satisfying a craving for a specific food affects their mood.

How a person's mood is affected may have something to do with why they are craving a particular food. Scientists speculate there are a number of reasons people may experience food cravings.

- They may feel overwhelming urges for specific foods because they forbid themselves to eat the foods. This frequently occurs with weight-conscious people.
- Pure pleasure, derived from the tastes and textures of certain foods, may be the driving factors. It has nothing to do with nutrients or physiologic drives.
- Our feelings about specific foods may drive cravings. If we are accustomed to eating certain foods at pleasant times during our lives, such as during the holidays, the foods may hold special attraction and make us feel better while we are eating them. If Mom comforted you with hot chocolate or ice cream, your desire for these foods during stressful times may reach stronger proportions.
- People may actually crave carbohdrates for their potential calming effect.

Carbohydrate Cravings and PMS

Scientists question why some women report carbohydrate cravings prior to menstruation, when they have premenstrual syndrome, and during pregnancy. Theories involve changes in the way food tastes to women, the increase in energy expenditure just prior to menstruation, and the potential effects of serotonin. Pregnant women frequently have very specific cravings, but why they crave the foods is unknown.

The Bottom Line

In the absence of a clear understanding of the effect of food and carbohydrates on mood, your best bet is to eat a well-balanced diet that includes the foods you like, even those you crave. That includes "indulgent" or "comfort" foods, which can be part of a healthy eating plan. If you try to avoid such foods when you crave them, you may set yourself up for overeating them in the long run. In this sense, favorite foods that conventional wisdom may say are "bad" for you may actually be very good for you. In moderation, they can help you stay with a well-balanced diet.

Chapter Seven

The Big Picture

EATING A LOW-FAT diet that contains plenty of carbohydrate-rich foods, such as whole-grain breads, cereals, pasta, rice, beans, fruits and vegetables, is the best dietary strategy for living a healthy life!

These foods form the foundation of the Food Guide Pyramid, the blueprint for healthy eating. The Food Guide Pyramid is based on the *Dietary Guidelines for Americans,* which provide advice for healthy Americans 2 years of age and over about food choices that promote health and prevent disease.

The *Dietary Guidelines* (see the following page) have featured advice to eat plenty of grain products, vegetables, and fruits since they were first issued back in 1980. However, Americans still fall short of getting the recommended amount of these important foods. A 1997 survey conducted by the Gallup organization showed only 12 percent of people (about one out of 10) reported eating the Food Guide Pyramid recommended 6 to 11 servings of grain foods a day. In contrast, over three-fourths of people *think* they are getting enough from this food group! And despite a several-year nationwide campaign to increase our consumption of fruits and vegetables, we still fall short in that area, too.

If you're among the majority of Americans who don't eat as many carbohydrate-rich foods as you need, you've got the information you need to make a change for the better in this book.

Dietary Guidelines for Americans

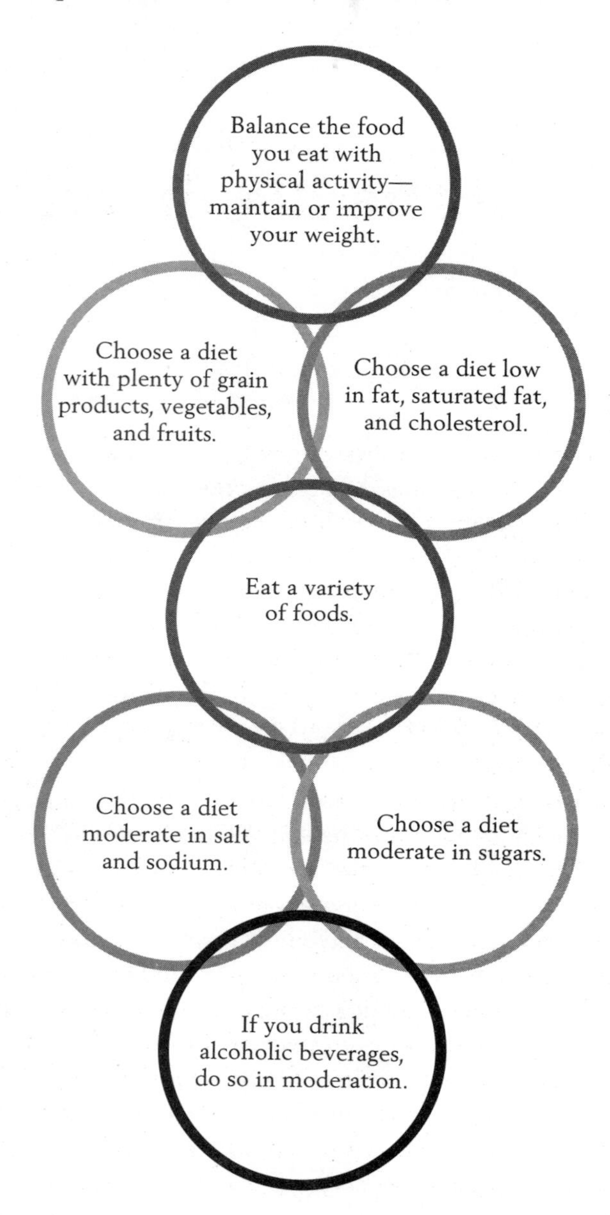

Source: U.S. Department of Agriculture, U.S. Department of Health and Human Services, 1995.

Start With the Food Guide Pyramid

The Food Guide Pyramid is an outline of what to eat each day. Following the pyramid recommendations will help you eat plenty of carbohydrate-rich foods, while keeping your intake of fats and sweets at a minimum.

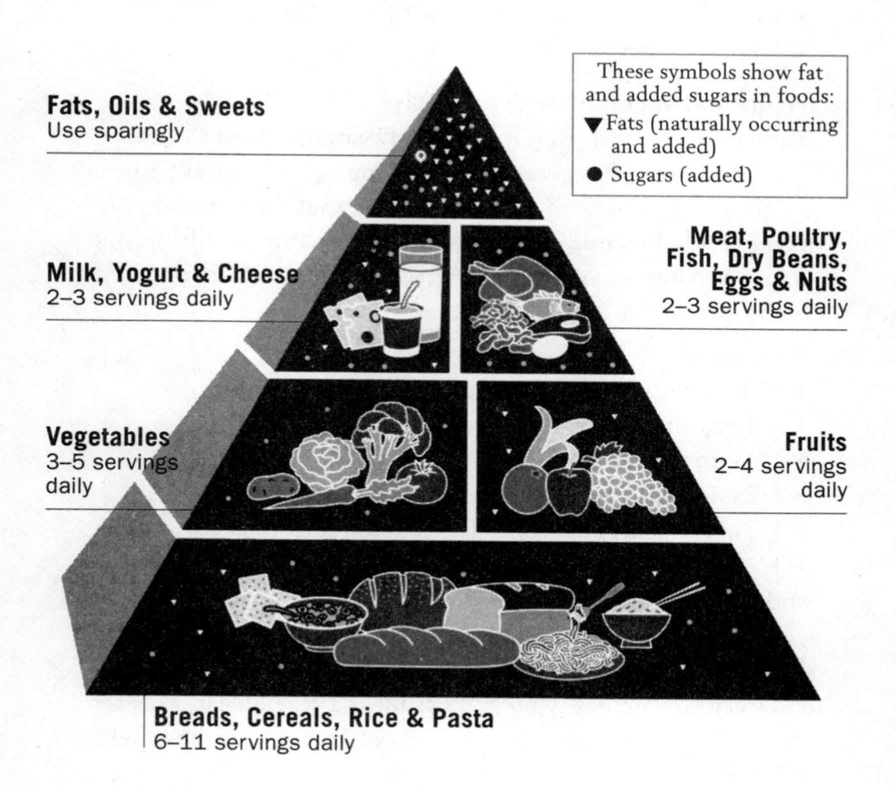

The Food Guide Pyramid illustrates the research-based food guidance system developed by the U.S. Department of Agriculture and supported by the Department of Health and Human Services.

6 to 11 Grain Servings A Day? You Must Be Kidding!

Do 6 to 11 servings of grain foods each day seem a bit much? It's really not when you consider the actual size of a serving. Take a look at just how big (or small!) a serving size is of these popular grain foods.

A serving is:

- ➤ 1 slice of bread
- ➤ 1 ounce ready-to-eat cereal
- ➤ 1/2 regular bagel
- ➤ 1/2 cup cooked pasta or rice
- ➤ 1 slice angel food cake
- ➤ 1 small muffin
- ➤ 1 small tortilla
- ➤ 1 4-inch pancake
- ➤ 1 slice of pizza
- ➤ 4 saltines or graham crackers
- ➤ 3 fig bar cookies
- ➤ 1 waffle

Fruits and Vegetables: Try 5 a Day!

Surveys show most people are already eating at least three servings of fruits and vegetables a day. Upping your intake to the minimum of five or more shouldn't be that far a stretch. The Food Guide Pyramid recommends 2 to 4 servings of fruits and 3 to 5 servings of vegetables. What's considered a fruit or vegetable serving?

A serving is:

- ➤ 1 medium fruit or 1/2 cup of small or cut-up fruit
- ➤ 3/4 cup of 100% fruit or vegetable juice
- ➤ 1/4 cup dried fruit
- ➤ 1/2 cup raw or cooked vegetables
- ➤ 1 cup raw leafy vegetables (such as lettuce, spinach)
- ➤ 1/2 cup cooked beans or peas (such as lentils, pinto beans, and kidney beans)

Eating More Grain Foods, Fruits, and Vegetables

How can you increase your grain, fruit, and vegetable intake each day, while keeping your fat intake low? Try some of these helpful tips.

Eat more grain foods:

- ➤ Start the day with ready-to-eat cereal with nonfat milk or whole wheat toast with jelly or jam.
- ➤ At lunch or dinner, put starch center stage. Try pasta with tomato sauce or rice with stir-fried beef and vegetables. Or, try cereal for a quick meal.
- ➤ For a low-fat snack option, try pretzels, rice cakes, air-popped popcorn, and of course, fresh fruit!

When you're at the supermarket, keep these shoppings tips in mind:

- Look for whole grain breakfast cereals, such as oatmeal and shredded wheat.
- Look for breads with whole wheat flour, stone-ground whole wheat flour or 100 percent whole wheat flour listed first on the ingredient list.
- Add variety as you choose low-fat breads—try bagels, tortillas, English muffins, pita bread sourdough, white, oatmeal, Italian, and French breads.
- Fill your shopping cart with a variety of grains—pasta, rice, bulgur, quinoa, and barley.
- If you buy packaged rice and pasta dishes, eliminate or cut back on the amount of oil or margarine called for in the directions.
- Buy baked versus fried potato and tortilla chips. Look for naturally low-fat salsas or reduced-fat dips to accompany them.
- Look for reduced-fat versions of your favorite snack crackers. Try naturally low-fat crackers, such as saltines, rye crackers, crispbreads, melba toasts and rounds, zwieback, and bread sticks.
- Look for reduced-fat versions of your favorite cookies. Try cookies that are naturally low in fat, such as fig and other fruit bars, gingersnaps, graham crackers, vanilla wafers, and animal crackers.
- Buy reduced-fat cake mixes or angel food cake mixes for a low-fat dessert. Top with fresh fruit or a fat free glaze.

Eat more fruits and vegetables.

Fruits and vegetables come in all sizes, shapes, and a lot of colors. They are available fresh, frozen, canned, and dried. And, they all count towards getting your 5 A Day. These days, the only limit to your choices is your imagination. Try some of these helpful tips.

- Grab a piece of fruit or two as a snack. Try dried fruit for a hardy take-it-with you snack that holds up well in your purse or briefcase.

- Thirsty? Just pour 100% fruit juice into a glass, add some ice, and enjoy.
- Try fresh raw vegetables for a crunchy snack.
- Steam or stir-fry vegetables for a full-flavor accompaniment to a meal.
- Add vegetables to chili, omelettes, pizza, stir-fried dishes, pastas, or stews.
- Add cooked black beans, corn, or garbanzo beans (chickpeas) to salads or casseroles.

At the supermarket:

- Fill your shopping cart with a variety of seasonal fresh fruits and vegetables.
- Buy canned and frozen fruits and vegetables to make sure you always have plenty on hand.
- Buy precut, packaged vegetables in the produce section to make quick side dishes and salads. Or, create a salad from the supermarket salad bar.
- Buy a variety of dried fruits such as raisins, prunes, apricots, and dates.
- Buy canned or frozen fruits with no sugar added and vegetables without butter, cream, or cheese sauces.
- Buy only 100% fruit juice. Punches, ades, and most fruit drinks contain only a little juice and lots of added sugar.
- Buy a variety of canned and dried beans and peas, such as pinto beans and lentils.
- Look for broth-based soups with lots of legumes and other vegetables—these contain considerably less fat than cream-based soups.

Putting It All Together

Now that you have plenty of information about carbohydrates, there's no better time than the present to get started with plans for healthy living. The sooner you invest in your health, the greater the benefit.

Remember that the big picture of healthy eating means getting the right balance of different nutrients. No one food (or nutrient) supplies everything you need. As shown in the Food Guide Pyramid, eating a wide variety of low-fat grain products, fruits, and vegetables (including beans and peas) is a great first step to begin meeting your nutritional needs.

The following sample two-day menu will show you how easy it is to eat a high-carbohydrate diet. Enjoy!

Menu #I

Calories: 1,728

Carbohydrate: 249 grams (58%)

Protein: 83 grams (19%)

Fat: 48 grams (25%)

Breakfast

- 6 ounces orange juice
- 1 cup whole grain oat cereal topped with
- 1/2 small, sliced banana
- 1 cup skim milk
- 1 slice of whole wheat toast
- 2 teaspoons soft margarine
- Coffee or Tea

Lunch

- 1 small hamburger (2-ounce patty) with bun, catsup, onion, pickles, tomato, lettuce
- Medium apple
- Water

Afternoon Snack

- 2 oatmeal cookies
- 3/4 cup seedless grapes
- 1 cup skim milk

Dinner

- Chicken cacciatore (3 ounces skinless chicken breast with 1/2 cup stewed tomatoes)
- 1/2 cup brown rice
- 1 1/2 cups mixed green salad (romaine lettuce, spinach, green onions, and cucumbers)
- 1 tablespoon Italian dressing
- Whole wheat dinner roll
- 1 cup skim milk
- 1/2 cup vanilla ice cream
- Water

Menu #2

Calories: 2,133

Carbohydrate: 310 grams (58%)

Protein: 88 grams (17%)

Fat: 65 grams (27%)

Breakfast

- 1/4 medium cantaloupe
- 1 large egg, soft-cooked
- 1 medium bran muffin
- 1 teaspoon soft margarine
- 1 tablespoon grape jelly
- 1 cup skim milk
- Coffee or tea

Lunch

- Ham and cheese sandwich (1 ounce lean ham, 1 ounce swiss cheese, 2 slices rye bread, 2 teaspoons mayonnaise, and lettuce & tomato)
- Carrot sticks and green pepper slices
- 1 small bag (1 ounce) potato chips
- Iced tea

Snack

- 4 graham crackers
- Fresh orange

Dinner

- Flounder florentine (3 ounces flounder, 1/4 cup spinach, 1/4 cup skim milk, 2 teaspoons Parmesan cheese)
- Baked potato
- 2 tablespoons reduced-fat sour cream
- 1/2 cup green peas
- 2 slices whole wheat bread
- 1 piece chocolate cake topped with 1/2 cup vanilla nonfat frozen yogurt

Appendix 1

Carbohydrate-Rich Foods

	Serving Size (1)*	Calories**	CHO (g)
Bread, cereal, rice and pasta			
bagel	1/2	80	15
biscuit (2-inch)	1	125	15
blueberry muffin	1 medium	110	20
bread	1 slice	65	10
bread sticks	2 4 1/2-inch sticks	40	5
bun (hamburger or hot dog)	1/2	60	10
cereal (ready-to-eat)	1 oz.	110	25
cereal (cooked)	1/2 cup	75	15
corn bread	1 2-inch square	145	20
English muffin	1/2	65	15
graham crackers	2 squares	55	10
pasta (cooked)	1/2 cup	100	20
pancakes (4-inch)	1	75	15
pita	1/2	60	15
popcorn (air popped, plain)	3 cups	90	20
pretzels	1 oz.	110	20
rice (cooked)	1/2 cup	105	20
saltine crackers	4	55	10
tortilla (flour, 6-inch)	1	75	15
waffle (4-inch)	1	65	10

*All serving sizes correspond to The Food Guide Pyramid serving sizes developed by the USDA. U. S. Department of Agriculture Human Nutrition Information Service, *The Food Guide Pyramid,* Home and Garden Bulletin No. 252, Washington, D.C., 1992.

**Energy, carbohydrate, and fiber values determined from University of Minnesota Nutrition Data System Version 2.9/11/26.

	Serving Size (1)*	Calories**	CHO (g)
Other baked goods			
angel food cake	1 slice	140	30
animal crackers	5	60	10
chocolate frosted cake	1 slice	460	75
fig bars	2	115	25
granola bar	1	105	15
raisin cookie	2 medium	120	20
Combination foods			
bean burrito	1	240	40
pizza (cheese)	1 slice	310	35
Fruits			
apple	1 medium	80	20
apple juice	3/4 cup	85	20
applesauce	1/2 cup	95	25
banana	1 medium	105	25
blueberries	1/2 cup	40	10
cantaloupe	1/4 medium	80	20
cherries (raw)	10	50	10
dates (dried)	5	115	30
fruit cocktail	1/2 cup	55	15
grape juice	3/4 cup	115	30
grapefruit	1/2 medium	40	10
grapes	1/2 cup	55	15
kiwi	1 medium	45	10
orange	1 medium	60	15
orange juice	3/4 cup	85	20
peach	1 medium	35	10
pear	1 medium	100	25
pineapple	1/2 cup	100	25
prunes (dried)	5	100	25
raisins	1/4 cup	115	30
raspberries	1/2 cup	30	5
strawberries	1/2 cup	25	5
watermelon	1/2 cup	25	5

	Serving Size (1)*	Calories**	CHO (g)
Vegetables			
carrot	1 medium	30	5
corn	1/2 cup	65	15
lima beans	1/2 cup	90	15
potato (baked, plain)	1 medium	135	30
sweet potato (baked, plain)	1 medium	115	30
Milk, yogurt and cheese			
frozen yogurt (low fat)	1 cup	200	35
fruit flavored yogurt (low fat)	1 cup	250	45
milk (1 %)	1 cup	100	10
milk (nonfat)	1 cup	85	10
pudding (made with nonfat milk)	1 cup	240	50

Appendix 2

Fiber-Rich Foods

	Serving Size (1)*	Calories**	Fiber (g)**,†		
			Total	Soluble	Insoluble
Fruits					
apple	1 medium	80	3.7	1.0	2.8
applesauce	1/2 cup	95	1.5	0.5	1.1
apple juice	3/4 cup	85	0.2	0.1	0.1
banana	1 medium	105	2.7	0.7	2.1
blueberries	1/2 cup	40	2.0	0.2	1.7
cantaloupe	1/4 medium	80	1.9	0.5	1.4
cherries (raw)	10	50	1.6	0.5	1.1
dates (dried)	5	115	3.1	0.5	2.6
figs (dried)	3	145	5.2	2.2	3.0
fruit cocktail	1/2 cup	55	1.3	0.7	0.6
grapefruit	1/2 medium	40	1.4	1.2	0.3
grapes	1/2 cup	55	0.8	0.3	0.5
grape juice	3/4 cup	115	0.2	0.2	0.0
kiwi	1 medium	45	2.6	0.6	2.0
orange	1 medium	60	3.1	1.8	1.3
orange juice	3/4 cup	85	0.4	0.2	0.2
prunes (dried)	5	100	3.0	1.6	1.4
peach	1 medium	35	1.7	0.7	1.0
pineapple	1/2 cup	100	0.9	0.3	0.6
pear	1 medium	100	4.0	2.2	1.8

*All serving sizes correspond to The Food Guide Pyramid serving sizes developed by the USDA. U. S. Department of Agriculture Human Nutrition Information Service, *The Food Guide Pyramid,* Home and Garden Bulletin No. 252, Washington, D.C., 1992.

**Energy, carbohydrate and fiber values determined from University of Minnesota Nutrition Data System Version 2.9/11/26.

†TF = total dietary fiber, SF = soluble fiber, IF = insoluble fiber. The sum of soluble and insoluble fiber values may not be equivalent to the total dietary fiber due to rounding.

	Serving Size (1)*	Calories**	Fiber (g)**,†		
			Total	Soluble	Insoluble
raisins	1/4 cup	115	1.6	0.4	1.1
raspberries	1/2 cup	30	4.2	0.4	3.8
strawberries	1/2 cup	25	1.8	0.5	1.3
watermelon	1/2 cup	25	0.4	0.2	0.2
Vegetables					
Cooked					
asparagus	1/2 cup	25	1.8	0.7	1.1
broccoli	1/2 cup	25	2.8	1.4	1.4
Brussel sprouts	1/2 cup	35	3.8	1.9	1.9
corn	1/2 cup	65	2.0	0.3	1.7
green beans	1/2 cup	20	2.0	0.9	1.2
peas	1/2 cup	60	4.4	1.3	3.1
potato (mashed)	1/2 cup	130	1.6	0.9	0.7
potato (baked, plain, with skin)	1 medium	135	2.9	1.2	1.7
spinach	1/2 cup	25	2.9	0.6	2.3
sweet potato (baked, plain)	1 medium	115	3.4	1.3	2.2
zucchini	1/2 cup	15	1.3	0.5	0.7
Raw					
carrot	1 medium	30	2.2	1.1	1.1
celery	1 stalk	5	0.7	0.2	0.4
cucumber (sliced)	1/2 cup	5	0.4	0.1	0.3
lettuce, Romaine	1 cup	5	0.7	0.2	0.5
mushrooms (sliced)	1/2 cup	10	0.4	0.1	0.4
spinach	1 cup	10	1.5	0.5	1.1
tomato	1 medium	25	1.4	0.1	1.2
Legumes					
Cooked					
baked beans (vegetarian)	1/2 cup	125	6.3	2.1	4.2
garbanzo beans	1/2 cup	135	4.4	1.3	3.0
kidney beans	1/2 cup	115	5.7	2.9	2.9
lentils	1/2 cup	115	7.8	0.6	7.2
navy beans	1/2 cup	130	6.5	2.2	4.3
soybeans	1/2 cup	155	5.4	2.4	3.0
Breads, grains and pasta					
barley	1/2 cup	95	4.3	0.9	3.3
bagel	1/2	80	0.6	0.3	0.3
bread, whole wheat	1 slice	70	2.0	0.3	1.6
bread, pumpernickel	1 slice	65	1.6	0.8	0.8
bread sticks	2 4 1/2-inch sticks	40	1.1	0.2	0.9
bread, French	1 slice	7	0.8	0.5	0.3
bread, white	1 slice	65	0.6	0.3	0.3

	Serving Size (1)*	Calories**	Fiber (g)**,†		
			Total	Soluble	Insoluble
bun (hamburger or hot dog)	1/2	60	0.6	0.2	0.4
pasta (cooked)	1/2 cup	100	0.9	0.4	0.6
pita	1/2	60	0.6	0.3	0.3
rice, brown (cooked)	1/2 cup	110	1.8	0.1	1.6
rice, white (cooked)	1/2 cup	105	0.3	0.1	0.2
Breakfast cereals					
100% bran	1 oz.	70	9.7	0.8	9.0
bran flakes	1 oz.	95	4.7	0.5	4.3
corn flakes	1 oz.	110	1.0	0.1	1.0
granola	1 oz.	135	2.4	1.4	1.1
oatmeal (cooked)	1/2 cup	75	2.0	0.9	1.1
puffed rice	1 oz.	115	0.5	0.1	0.4
raisin bran	1 oz.	85	2.8	0.7	2.1
whole grain oats cereal	1 oz.	105	2.9	1.3	1.5
Snacks					
corn chips	1 oz.	155	1.4	0.0	1.4
hummus dip	2 tablespoons	80	2.0	0.6	1.4
popcorn (air popped, plain)	3 cups	90	3.6	0.1	3.5
peanuts (dry roasted)	1/4 cup	210	3.4	0.7	2.7
pretzels	1 oz.	110	0.9	0.3	0.7
sunflower seeds	1/4 cup	180	3.4	0.7	2.7
walnuts	1/4 cup	160	1.2	0.4	0.8
Added Ingredients					
gums	0.1 oz.	5	2.8	2.8	0.0
flax seeds	1 tablespoon	35	1.6	0.8	0.7
oat bran	1 tablespoon	15	0.9	0.4	0.5
psyllium	1 tablespoon	5	6.0	4.8	1.2
rice bran	1 tablespoon	15	1.1	0.1	1.0
seaweed	1 tablespoon	1	0.4	0.4	0.0
wheat bran	1 tablespoon	10	1.6	0.1	1.5
wheat germ	1 tablespoon	25	0.9	0.2	0.7

Index